FISHERIES

2

PRECAUTIONARY APPROACH TO CAPTURE FISHERIES AND SPECIES INTRODUCTIONS

Elaborated by the
Technical Consultation on the Precautionary Approach
to Capture Fisheries (Including Species Introductions)
Lysekil, Sweden, 6-13 June 1995

FOOD AND AGRICULTURE ORGANIZATION OF THE UNITED NATIONS
Rome, 1996

M-40
ISBN 92-5-103915-1

PREPARATION OF THIS DOCUMENT

These Guidelines were originally drafted by the Technical Consultation on the Precautionary Approach to Capture Fisheries (Including Species Introductions) organized by the Government of Sweden in cooperation with FAO in the Fisheries Laboratory of Lysekil (Sweden, 6-13 June 1995) with the participation of the following experts:

Devin Bartley, Åsmund Bjordal, John F. Caddy, Kee-Chai Chong, Engel J. De Boer, William De la Mare, Chris Francis, Serge Garcia, Henrik Gislasson, Olle Hagström, Ray Hilborn, Mikael Hildén, Daniel Huppert, Eskild Kirkegaard, Geoffrey Kirkwood, Christian Lévêque, Armin Lindquist, Jordi Lleonart, Alec MacCall, Jean-Jacques Maguire, Robin Mahon, Dan Minchin, Randall Peterman, John Pope, Andrew Rosenberg, Keith Sainsbury, Juan Carlos Seijo, Fred Serchuk, Ross Shotton, Michael Sissenwine, Tony Smith, Ziro Suzuki, Jan Thulin, Per Wramner.

The Code of Practice on the Introduction and Transfer of Marine Organisms (1994) is included in the Report as Annex A, by courtesy of the International Council for the Exploration of the Sea (ICES). The Guidelines for Preventing the Introduction of Unwanted Aquatic Organisms and Pathogens from Ships' Ballast Water and Sediment Discharges (1994) is included in the Report as Annex B by courtesy of the International Maritime Organization (IMO).

The Guidelines, originally published as FAO Fisheries Technical Paper, No. 350, Part 1, Rome, FAO, 1995 are reproduced in this new FAO Series (FAO Technical Guidelines for Responsible Fisheries) with minor editing.

The scientific contributions by experts attending the Technical Consultation in Lysekil, which represent a very useful corpus of additional information on the precautionary approach to fisheries have been published as FAO Fisheries Technical Paper, No. 350, Part 2, Rome, FAO, 1996.

Cover page: the logo of the Institute of Marine Research, Lysekil:

Bronze-age rock carving from the parish of Kville, Bohuslän, Sweden. From thousands of rock-carvings in western Sweden this is the only known scene showing fishing. Originally described by Åke Fredsjö, 1943: En fiskescen på en bohuslänsk hällristning - Göteborgs och Bohusläns Fornminnesförenings tidskrift 1943: 61-71. Later documentation by the same author in: Hällristningar i Kville härad i Bohuslän. Kville socken. Del 1 och 2. - Studier i nordisk arkeologi 14/15, Göteborg 1981, 303 pp., Pl. 158 II. Published by Fornminnesföreningen i Göteborg.

FAO.
Precautionary approach to capture fisheries and species introductions.
Elaborated by the Technical Consultation on the Precautionary Approach to
Capture Fisheries (Including Species Introductions). Lysekil, Sweden,
6-13 June 1995.
FAO Technical Guidelines for Responsible Fisheries. No. 2. Rome, FAO.
1996. 54p.

ABSTRACT

Starting from Principle 15 of the Rio Declaration (UNCED, 1992), the
document proposes a definition of the **precautionary approach to fisheries** as
well as an elaboration on the **burden of proof.** It also contains detailed
guidelines on how to conduct fishery management and research and how to
develop and transfer fishery technology in a context of uncertainty and
responsible fisheries.

Guidelines are also provided on species introduction, voluntary or accidental
(including through ballast water and sediment discharge), recognizing the
difficulty of ensuring a precautionary approach in relation to that issue. The
guidelines are aimed at the governments, fisheries authorities, fishery industry,
regional fishery management bodies, NGOs and other interested parties, in
order to: (a) raise their awareness about the need for precaution in fisheries, by
providing them with background information on the main issues and
implications, and (b) provide practical guidance on how to apply such
precaution.

Distribution:

All FAO Members and Associate Members
Participants
Institute of Marine Research, Lysekil, Sweden
Other interested Nations and International Organizations
FAO Fisheries Department
FAO Fishery Officers in FAO Regional Offices
Non-governmental Organizations

CONTENTS

1. From ancient times, fishing has been a major source of food for humanity and a provider of employment and economic benefits to those engaged in this activity. However, with increased knowledge and the dynamic development of fisheries it was realised that aquatic resources, although renewable, are not infinite and need to be properly managed, if their contribution to the nutritional, economic and social well-being of the growing world's population was to be sustained.

2. The adoption in 1982 of the United Nations Convention on the Law of the Sea provided a new framework for the better management of marine resources. The new legal regime of the oceans gave coastal States rights and responsibilities for the management and use of fishery resources within their EEZs which embrace some 90 percent of the world's marine fisheries.

3. In recent years, world fisheries have become a dynamically developing sector of the food industry and coastal States have striven to take advantage of their new opportunities by investing in modern fishing fleets and processing factories in response to growing international demand for fish and fishery products. It became clear, however, that many fisheries resources could not sustain an often uncontrolled increase of exploitation.

4. Clear signs of over-exploitation of important fish stocks, modifications of ecosystems, significant economic losses, and international conflicts on management and fish trade threatened the long-term sustainability of fisheries and the contribution of fisheries to food supply. Therefore the Nineteenth Session of the FAO Committee on Fisheries (COFI), held in March 1991, recommended that new approaches to fisheries management embracing conservation and environmental, as well as social and economic, considerations were urgently needed. FAO was asked to develop the concept of responsible fisheries and elaborate a Code of Conduct to foster its application.

5. Subsequently, the Government of Mexico, in collaboration with FAO, organised an International Conference on Responsible Fishing in Cancún, in May 1992. The Declaration of Cancún endorsed at that Conference was brought to the attention of the UNCED Rio Summit in June 1992, which supported the preparation of a Code of Conduct for Responsible Fisheries. The FAO Technical Consultation on High Seas Fishing, held in September 1992, further recommended the elaboration of a Code to address the issues regarding high seas fisheries.

6. The One Hundred and Second Session of the FAO Council, held in November 1992, discussed the elaboration of the Code, recommending that priority be given to high seas issues and requested that proposals for the Code be presented to the 1993 session of the Committee on Fisheries.

7.   The Twentieth Session of COFI, held in March 1993, examined in general the proposed framework and content for such a Code, including the elaboration of guidelines, and endorsed a timeframe for the further elaboration of the Code. It also requested FAO to prepare, on a "fast track" basis, as part of the Code, proposals to prevent reflagging of fishing vessels which affect conservation and management measures on the high seas. This resulted in the FAO Conference, at its Twenty-seventh Session in November 1993, adopting the Agreement to Promote Compliance with International Conservation and Management Measures by Fishing Vessels on the High Seas, which according to FAO Conference resolution 15/93 forms an integral part of the Code.

8.   The Code was formulated so as to be interpreted and applied in conformity with the relevant rules of international law, as reflected in the United Nations Convention on the Law of the Sea, 1982, as well as with the Agreement for the Implementation of the Provisions of the United Nations Convention on the Law of the Sea of 10 December 1982 Relating to the Conservation and Management of Straddling Fish Stocks and Highly Migratory Fish Stocks, 1995, and in the light of *inter alia* the 1992 Declaration of Cancún, the 1992 Rio Declaration on Environment and Development, in particular Chapter 17 of Agenda 21.

9.   The development of the Code was carried out by FAO in consultation and collaboration with relevant United Nations Agencies and other international organisations including non-governmental organisations.

10.   The Code of Conduct consists of five introductory articles: Nature and Scope; Objectives; Relationship with Other International Instruments; Implementation, Monitoring and Updating; and Special Requirements of Developing Countries. These introductory articles are followed by an article on General Principles which precedes the six thematic articles on: Fisheries Management, Fishing Operations, Aquaculture Development, Integration of Fisheries into Coastal Area Management, Post-Harvest Practices and Trade, and Fisheries Research. As already mentioned, the Agreement to Promote Compliance with International Conservation and Management Measures by Fishing Vessels on the High Seas forms an integral part of the Code.

11.   The Code is voluntary. However, certain parts of it are based on relevant rules of international law, as reflected in the United Nations Convention on the Law of the Sea of 10 December 1982. The Code also contains provisions that may be or have already been given binding effect by means of other obligatory legal instruments amongst the Parties, such as the Agreement to Promote Compliance with Conservation and Management Measures by Fishing Vessels on the High Seas, 1993.

12.   The Twenty-eighth Session of the Conference in Resolution 4/95 adopted the Code of Conduct for Responsible Fisheries on 31 October 1995. The same Resolution requested FAO *inter alia* to elaborate as appropriate technical guidelines in support of the implementation of the Code in collaboration with members and interested relevant organisations.

Sustainable development has been defined as *"the management and conservation of the natural resource base, and the orientation of technological and institutional change in such a manner as to ensure the attainment and continued satisfaction of human needs for present and future generations. Such development conserves land, water, plant genetic resources, is environmentally non-degrading, technologically appropriate, economically viable and socially acceptable."* (FAO Council, 94th Session, 1988).

Principle 15 of the Rio Declaration of the UN Conference on Environment and Development (Rio de Janeiro, 1992) states that *"In order to protect the environment, the precautionary approach shall be widely applied by States according to their capabilities. Where there are threats of serious or irreversible damage, lack of full scientific certainty shall be not used as a reason for postponing cost-effective measures to prevent environmental degradation."*

The General Principles and Article 6.5 of the FAO International Code of Conduct for Responsible Fisheries adopted by the FAO Conference in 1995, prescribe a precautionary approach to all fisheries, in all aquatic systems, and regardless of their jurisdictional nature, recognizing that most problems affecting the sector result from insufficiency of precaution in management regimes when faced with high levels of uncertainty.

The United Nations Conference on Straddling Fish Stocks and Highly Migratory Fish Stocks (New York, 1992-1995) developed a consensus on the need to introduce or strengthen the precautionary approach to fishery management, imbedding the concept in the draft text of its outcome, and outlining elements for its implementation.

Because uncertainty affects all elements of the fishery system in varying degrees, some degree of precaution is required at all levels of the system: in development planning, management, research, technology development and transfer, legal and institutional frameworks, fish capture and processing, fisheries enhancement and aquaculture.

The following guidelines have been developed by the Technical Consultation on the Precautionary Approach to Capture Fisheries (Lysekil, June 1995), for the governments, fisheries authorities, the fishing industry, regional fishery management bodies, NGOs, and others interested in fisheries, in order to: (a) raise their awareness about the need for precaution in fisheries, by providing them with background information on the main issues and implications, and (b) provide them with practical guidance on how to apply such precaution.

These Guidelines are preliminary and will be evaluated and revised as information accumulates through their implementation.

1. In line with its commitment towards environmental conservation and sustainable use of natural resources, the Government of Sweden (through its Ministry of Agriculture) decided to organize, in close cooperation with FAO, a Technical Consultation on the Precautionary Approach to Capture Fisheries (including Species Introductions). The meeting was hosted by the Institute of Marine Research, Lysekil, Sweden at the invitation of the Swedish National Board of Fisheries and was formally opened by her Excellency the Swedish Minister of Agriculture, Mrs Margareta Winberg.

2. The meeting was chaired by Per Wramner (Swedish National Board of Fisheries), assisted by Armin Lindquist (Swedish National Board of Fisheries) and Serge Garcia (FAO, Vice-Chairman). The participants attended in their personal capacity and were selected on the basis of their technical competence and level of expertise. Their deliberations were based on 7 background documents prepared specifically for the purpose, as well as on a number of other documents considered of relevance. Four Working Groups were established to discuss and prepare the sections of the draft guidelines related to: Research, Management, Technology and Species Introductions.

3. The Administrative Report of the meeting, including Agenda, List of Participants, background documentation, opening address by the Swedish Minister of Agriculture, and composition of Working Groups, has been published as *FAO Fisheries Report*, (527), 1995.

4. The following document first characterises the concept of precaution, then defines some of the important terms used in the Guidelines, and provides specific guidelines for management, research, technology development and transfer, and species introduction. Under each of these topics, a discussion is provided on its specific aspects, followed by specific guidance for implementation.

1. PRECAUTIONARY APPROACH AND BURDEN OF PROOF

5. Within the framework outlined in Article 15 of the UNCED Rio Declaration, the precautionary approach to fisheries recognises that changes in fisheries systems are only slowly reversible, difficult to control, not well understood, and subject to changing environment and human values.

6. The precautionary approach involves the application of prudent foresight. Taking account of the uncertainties in fisheries systems and the need to take action with incomplete knowledge, it requires, *inter alia:*

    a. consideration of the needs of future generations and avoidance of changes that are not potentially reversible;

    b. prior identification of undesirable outcomes and of measures that will avoid them or correct them promptly;

    c. that any necessary corrective measures are initiated without delay, and that they should achieve their purpose promptly, on a timescale not exceeding two or three decades;

    d. that where the likely impact of resource use is uncertain, priority should be given to conserving the productive capacity of the resource;

    e. that harvesting and processing capacity should be commensurate with estimated sustainable levels of resource, and that increases in capacity should be further contained when resource productivity is highly uncertain;

    f. all fishing activities must have prior management authorization and be subject to periodic review;

    g. an established legal and institutional framework for fishery management, within which management plans that implement the above points are instituted for each fishery, and

    h. appropriate placement of the burden of proof by adhering to the requirements above.

7. Key concepts in past discussions of the precautionary approach have been the burden of proof and the standard of proof (i.e., the responsibility for providing the relevant evidence and the criteria to be used to judge that evidence). Often, the precautionary approach has been taken as requiring that human actions are assumed to be harmful unless proven otherwise (reversal of the burden of proof). In regard to these concepts, it is recognised that:

a. all fishing activities have environmental impacts, and it is not appropriate to assume that these are negligible until proved otherwise;

b. although the precautionary approach to fisheries may require cessation of fishing activities that have potentially serious adverse impacts, it does not imply that no fishing can take place until all potential impacts have been assessed and found to be negligible;

c. the precautionary approach to fisheries requires that all fishing activities be subject to prior review and authorization; that a management plan be in place that clearly specifies management objectives and how impacts of fishing are to be assessed, monitored and addressed; and that specified interim management measures should apply to all fishing activities until such time as a management plan is in place, and

d. the standard of proof to be used in decisions regarding authorization of fishing activities should be commensurate with the potential risk to the resource, while also taking into account the expected benefits of the activities.

2. DEFINITIONS

8. **Decision Rule:** Specification of how pre-agreed management actions will respond to estimated or perceived states of nature.

9. **Fishery Technology:** The equipment and practices used for finding, harvesting, handling, processing and distributing aquatic resources and their products.

10. **Genetically Modified Organism:** An organism in which the genetic material has been altered anthropogenically by means of gene or cell technologies.

11. **Genetically selected organism:** An organism produced by selective breeding.

12. **Introduced Species:** Any species intentionally or accidentally transported and released by humans into an environment beyond its present range.

13. **Management procedure:** A description of the data to collect, how to analyze it, and how the analysis translates into actions.

14. **Risk:** The probability of something undesirable happening (note that when a technical definition in a decision theoretic framework is needed, it would be appropriate to use the terms "expected loss" or "average forecasted loss", not risk)

15. **States of Nature:** A description of a condition and dynamics of the resource and the fishery including parameters such as stock abundance, age structure, fishing mortality, the economic condition of the industry and the state of the environment.

16. **Statistical uncertainty**: Stochasticity or error from various sources as described using statistical methodology.

17. **Transferred species**: Any species intentionally or accidentally transported and released by humans into an environment inside its present range.

18. **Uncertainty**: The incompleteness of knowledge about the state or processes of nature.

3. PRECAUTIONARY APPROACH TO FISHERY MANAGEMENT

3.1 Introduction

19. Management according to the precautionary approach exercises prudent foresight to avoid unacceptable or undesirable situations, taking into account that changes in fisheries systems are only slowly reversible, difficult to control, not well understood, and subject to change in the environment and human values.

20. An important element of the precautionary approach is to establish legal or social management frameworks for all fisheries, which is not the current situation. At a minimum, such frameworks should establish rules controlling access to fisheries (e.g., all boats must be licensed), data reporting requirements, and processes for planning and implementing more comprehensive fishery management. Plans for management institutionalize prudent foresight that takes into account potential consequences of fishery development and events affecting it. Comprehensive plans for fisheries can take a long time to develop. For this reason the legal or social management framework should include interim measures that safeguard the resources until such plans are adopted.

21. The precautionary approach gives due concern to long-term effects in the specification of management objectives and in the development of management frameworks, procedures, and measures. The consequences of management and fishery development are evaluated to reduce the possibilities of changes that are not potentially reversible on a 2 to 3 decade time scale. Processes for determining acceptable changes and impacts are used to support the precautionary approach. Thus, a precautionary approach links fisheries management intimately with general environmental management.

22. Precautionary management involves explicit consideration of undesirable and potentially unacceptable outcomes and provides contingency and other plans to avoid or mitigate such outcomes. Undesirable or unacceptable outcomes include overexploitation of resources, overdevelopment of harvesting capacity, loss of biodiversity, major physical disturbances of sensitive biotopes, or social or economic dislocations. Undesirable conditions can also arise when a fishery is negatively influenced by other fisheries or other activities and when management fails to take action in the face of shifts in the external conditions affecting, for example, the productivity of the fish stocks.

23. The operational interpretations of precautionary management will depend on the context. Different interpretation may be appropriate depending on the scale of the fishing operations (artisanal or small-scale fisheries vs. highly capitalized and technologically advanced fisheries) and on the state of the exploited system (early stages of exploitation versus systems in a state of obvious overexploitation).

24. The precautionary approach is included in all stages of the management process. Thus, precaution should be identifiable in the different stages of management, from planning through implementation, enforcement and monitoring to re-evaluation. These issues are covered in the following paragraphs organized according to the different stages in a management process.

3.2 Management Planning

25. A precautionary approach to managing a fishery involves developing, within management strategies and plans, explicit consideration of precautionary actions that will be taken to avoid specific undesirable outcomes. As over-development of harvesting capacity is a common cause of undesirable outcomes, a management plan should include mechanisms to monitor and control that capacity. Consideration needs to be given to how uncertainty and ignorance are to be taken into account in developing and varying management measures. For all fisheries, plans should be developed or revised to incorporate precautionary elements. The plans, even where no additional precautionary elements are considered necessary, should be re-evaluated in accordance with the process outlined below. Where there are multiple fisheries, plans will also be required to implement precautionary approaches to their aggregate impact on the marine environment. The plans should consider time scales of at least two to three decades, or longer in the case of long-lived species.

26. To ensure broad acceptance, all stages of planning should involve consultation with the fishing industry, conservation groups, and other interested parties. Fisheries plans should also be coordinated with integrated coastal-area management plans. In order to identify a management plan that has broad acceptance, it is best to consider a range of alternatives, each of which has been developed and evaluated through the components set out below. The range of alternatives may differ in their basic approach or in detail. For example, a basic approach using total allowable catches (TACs) could be contrasted with one using effort controls. Variations in detail might involve different decision rules for the TACs.

Specifying management objectives

27. The first step is to identify the broad management objectives to be achieved. The management objectives need to consider both the manner in which the benefits from the fishery are to be realized, as well as the possible undesirable outcomes which are to be avoided. Broad objectives include considerations of long-term interests and the avoidance of irreversible or slowly reversible changes. Typically, the catches are to be as large as possible, so long as the probability of substantial stock depletion is below an acceptably low level and catches can be kept reasonably steady.

28. The general objectives could be taken as the starting point for setting the more specific objectives for a particular fishery. To be precautionary, priority should be accorded to restoration of already overfished stocks, to avoidance of overfishing, and to avoidance of excessive harvesting capacity. Objectives should also include restricting the environmental impacts of fishing to acceptable levels. Some examples are limiting or eliminating bycatch and incidental mortality of non-target species and containing the possible effects of some types of fishing gear on bottom communities.

Specifying operational targets and constraints

29. Targets identify the desired outcomes for the fishery. For example, these may take the form of a target fishing mortality, or a specified level of average abundance relative to the unfished state. In some cases, these targets are likely to be identical with those that would be specified for fisheries management, regardless of whether a precautionary approach was to be adopted. In other cases, targets may need to be adjusted to be precautionary, for example, by setting the target fishing mortality lower than F_{MSY}.

30. The operational constraints explicitly define the undesirable outcomes that are to be avoided. For example, to avoid the risk of declining recruitment, a minimum spawning stock biomass, range of ages, or geographic range could be set to define safe limits within which the stock should be maintained with a specified high probability. Specific limits may also be required to deal with ecosystem effects, with bycatches and with other side-effects of the fishery.

31. Operational targets and constraints should be expressed in measurable terms such as target reference points and limit reference points (refer to FAO documents). The details of what can be measured will often vary with different species and fisheries, and so the operational targets and constraints will need to be expressed in ways that take this into account. The specification of operational targets and constraints cannot be separated from consideration of the types of data and methods that can be used to assess the status of the stocks. In all cases attention should be given to the rate at which targets are approached so as to avoid overshooting them and hence violating the constraints.

Specifying the procedure to apply and adjust management measures

32. A management plan must indicate which management measures are to be applied, and the circumstances under which the measures are to be varied. This should involve the formulation of decision rules, which specify in advance what action should be taken when specified deviations from the operational targets and constraints are observed. The specification should include minimum data requirements for the types of assessment methods to be used for decision-making.

33. Precautionary management measures listed below under "Examples of Precautionary Measures" could be included in the plan. To be precautionary, decision rules are required for responding to unexpected or unpredictable events with minimum delay.

All foreseeable contingencies should be considered when developing the plan. For example, plans should include explicit effort-reduction measures that apply in response to unpredicted, marked decline in recruitment.

34.   It is highly desirable that the procedure makes regular small adjustments to the management measures so as to maintain acceptably low levels of probability that the constraints are violated. It is not always possible to simultaneously attain a target (desired outcome) for a fishery and respect constraints designed to prevent undesirable outcomes. For example, a specified target fishing mortality such as F_{MSY} may reduce the spawning stock biomass to a level near the levels where there should be a precautionary constraint designed to avoid the probability of declining recruitment. If, for example, the constraint is to maintain spawning stock biomass above 30% of the average unfished level with high probability, then a F_{MSY} target that would reduce the spawning stock biomass to 35% of the unfished level could have a too high probability of violating the constraint. Precautionary management must adjust targets to be consistent with the constraints.

Prospective evaluation

35.   A precautionary approach requires that the feasibility and reliability of the management options be evaluated. A management plan should not be accepted until it has been shown to perform effectively in terms of its ability to avoid undesirable outcomes. The evaluation can be used to determine whether the data and assessment methods available for management are sufficient to meet the management objectives. The evaluation should attempt to determine if the management plan is robust to both statistical uncertainty and to incomplete knowledge on factors such as uncertain stock identity and abundance, stock dynamics, and the effects of environmental variability and trends. As well, evaluations should consider the dynamic behaviour of the harvesting sector and managers' ability to change harvest levels.

36.   For economically valuable fisheries, and where substantial scientific expertise is available, there will usually be substantial benefits from employing powerful evaluation techniques such as simulation modelling. Such analyses will often reveal which sources of uncertainty are critical to achieving satisfactory results for the various objectives. The evaluation will also need to take into account the practicality of implementing, and securing compliance with, the range of management measures included in the plan.

37.   For small fisheries and artisanal fisheries, computationally intensive management analyses are often not possible or cost-effective. In such cases, management measures will probably not depend on quantitative analyses, but rather on assessing the practicality of ensuring that the precautionary measures are accepted and observed by the fishing community. An example would be closing certain areas to fishing to protect a sufficient proportion of the stock. Another example would be to establish a community-based fisheries management system. This would decentralize fisheries management authority to resource users and could reduce the cost of fisheries management and enforcement. Other examples of simple precautionary measures applicable to such fisheries are given in the section "Examples of Precautionary Measures" below.

38.   If management options are found to be inadequate with respect to precaution, then one or more of the following aspects can be modified and then re-evaluated until the management system is judged to be adequate. These aspects may include:

    a.   modification of the operational targets and constraints;
    b.   re-specification of the procedure to apply management measures;
    c.   further research to reduce critical uncertainties, or
    d.   consideration of more powerful assessment and monitoring methods.

3.3 Implementation, Monitoring, and Enforcement

39.   Management plan implementation puts in place all planned decision rules. This involves the practical interpretation of objectives and procedures, and the implementation of detailed instructions for compliance, monitoring of the fishery, and enforcement tactics. Elements of the implementation phase include: stock assessments, rule setting, economic assessments, and communication of decisions and rationale to the public and fishing industry. Because the public and industry are more inclined to understand and support measures on which they are consulted, public participation in the implementation phase is important. Peer review of stock assessments and a transparent process help to guard against error, which is essential to effective implementation of the planned measures. Independent auditing of the monitoring procedures should also be a regular feature of the management system. The effect of the measures on compliance should be studied specifically.

40.   Monitoring of a fishery involves collection of all information relevant to ensuring that the plan is being executed and that it is achieving the desired results. In particular, data are needed to determine whether that precautionary decision rules are being violated. A precautionary approach to monitoring will use many and various sources of information, including environmental and socio-economic data.

41.   Precautionary monitoring of fishing should seek to detect and observe a variety of ancillary impacts, e.g., environmental changes, fish habitat degradation, and effects on birds, mammals and other biota. This monitoring function could use information from fishing participants, indigenous people, and other public groups, and have appropriate procedures to process and analyze this information.

42.   In a precautionary management system, contingency rules should be implemented to ensure compliance with operational targets and constraints in the face of major adverse events with low probability. There should also be mechanisms for revising targets and constraints in the light of unexpected events.

43.   A precautionary system of enforcement and the penalties for non-compliance should have the flexibility for prompt action by redeployment of monitoring and enforcement resources. For example, the first signs of bycatch problems should be followed by more extensive sampling in problem areas according to an agreed procedure or enhanced surveillance of the fishery. In the case of emergency, it should be possible to rapidly modify regulations.

3.4 Re-evaluation of Management Systems

44. The level of precaution in the management system needs to be re-assessed periodically. This includes: (1) the degree of precaution in the objectives, operational targets and constraints in relation to observed changes in the fishery and the environment, (2) the use of scientific information and other information in the management process, (3) the applicability of the contingency plans for unexpected conditions, and (4) auditing of all procedures in the fisheries management system. Special re-evaluations should be initiated as soon as it becomes apparent that the fishery inadvertently violates the limit reference points established in the plan.

3.5 Implementation guidelines

45. There are several precautionary measures that fisheries management agencies should take in order to avoid undesirable or unacceptable outcomes in the development of fisheries. Some of these will apply to all types of fisheries, whereas others will be useful only in specific situations such as overexploited fisheries. For illustrative purposes, we list precautionary measures for four typical situations: (1) new or developing fisheries; (2) overutilized fisheries; (3) fully utilized fisheries; and (4) traditional or artisanal fisheries.

46. The listed measures could be included in comprehensive fisheries plans, but could also be used in the interim for immediate precautionary action. An example interim measure in the case of new fisheries would be a conservative cap on fishing effort. In overutilized fisheries an interim measure would be a rapid reduction of fishing mortality. Once various proposed management plans have been evaluated by the methods discussed above, the approved plan can replace the interim action.

New or developing fisheries

47. Some of the precautionary measures listed below for new or still-developing fisheries will also apply to fully utilized, overutilized, or artisanal fisheries, as described later. Most of these recommendations also apply to existing fisheries that are not yet managed:

a. always control access to the fishery early, before problems appear. An open access fishery is not precautionary;

b. immediately put a conservative cap (or default level) on both fishing capacity and the total fishing mortality rate. This could be achieved by limiting effort or total allowable catch. As well, attention should be paid to preventing excessive investment in the processing sector. The conservative caps should remain in place until analyses of data justify an increase in fishing effort or fishing mortality. The objective is to prevent that the development of the fleet's fishing power and capacity outpaces the ability of management to understand the effect of existing fishing effort;

c.   build in flexibility so that it is feasible to phase vessels out of the fleet, if this becomes necessary. To avoid new investments in fishing capacity, temporarily license vessels from another fishery;

d.   to limit risks to the resource and the environment, use area closures, which are relatively quick to implement and are easily enforceable. Closures provide refuges for fish stocks, protect habitat, and provide areas for comparison with fished areas;

e.   establish precautionary, preliminary biological limit reference points (e.g., spawning stock biomass less than 50% of the initial biomass) in the planning stage as described above;

f.   encourage fishing in a responsible manner to ensure long-term persistence of a productive stock or other parts of the ecosystem. For instance, encourage voluntary agreements on conduct in the fishery through co-management, community management, or some form of tenure of fishing rights;

g.   encourage development of fisheries that are economically viable without long-term subsidies;

h.   establish a data collection and reporting system for new fisheries early in their development;

i.   immediately start a research programme on the stock and fisheries, including the response of individual vessels to regulations. When issuing a fishing license, require a vessel to report detailed information, including standard biological data and economic information, and

j.   take advantage of any opportunities for setting up experimental situations to generate information on the resources. This could be done by contrasting different harvesting strategies on subpopulations, for instance.

Overutilized fisheries

48.   Most of the above recommendations also apply to fish stocks that are already overutilized, but in addition, special precautionary measures need to be taken for such stocks. These are:

a.   immediately limit access to the fishery and put a cap on a further increase in fishing capacity and fishing mortality rate;

b.   establish a recovery plan that will rebuild the stock over a specific time period with reasonable certainty. This will include several of the components below;

c. reduce fishing mortality rates long enough to allow rebuilding of the spawning stock. If possible, take immediate short-term action even on the basis of circumstantial evidence about the effectiveness of a particular measure. In some cases this can be accomplished by entirely closing some areas to fishing;

d. when there is a good year class, give priority to using the recruits to rebuild the stock rather than increasing the allowable harvest;

e. reduce fishing capacity to avoid recurrence of over-utilization. Remove excessive fishing capacity from the fishery; do not provide subsidies or tax incentives to maintain fishing capacity. If necessary, develop mechanisms to eliminate some fishing effort;

f. alternatively, allow vessels to move from an overutilized fishery into another fishery, as long as the pressure from this redeployment does not jeopardize the fishery that the vessels are moving into;

g. do not use artificial propagation as a substitute for the precautionary measures listed above;

h. in the management plan, establish biological reference points to define recovery, using measures of stock status, such as spawning stock biomass, spatial distribution, age structure, or recruitment, and

i. for species where it is possible, closely monitor the productivity and total area of required habitat to provide another indicator of when management action is needed.

Fully utilized fisheries

49. These are fisheries that are heavily harvested but not yet overexploited. Regulatory agencies must particularly watch for signs that the population is becoming overexploited. While some precautionary measures from the above lists apply here, additional measures to take in this situation are:

a. ensure that there are means to effectively keep fishing mortality rate and fishing capacity at the existing level;

b. there are many "early-warning signs" that a stock is becoming overutilized (e.g., age structure of the spawners shifting to an unusually high proportion of young fish, shrinking spatial distribution of the stock or species composition in the catch). These warning signs should trigger investigative action according to prespecified procedures while interim management actions are taken, as noted below;

c.   when precautionary or limit reference points are approached closely, prespecified measures should be taken immediately to ensure that they will not be exceeded (i.e., do not wait until violation of a limit point is imminent to start deciding what to do about it);

d.   if limit reference points are exceeded, recovery plans should be implemented immediately to restore the stock. The recommendations for overutilized stocks described above should then be implemented;

e.   to prevent excessively reducing the reproductive capacity of a population, avoid harvesting immature fish, unless there is strong protection of the spawning stock. For example, if immature fish exceed a specified percentage of the catch, close the local area to all harvesting.

Traditional or artisanal fisheries

50.   These are low-technology fisheries carried out by large numbers of small vessels, often where there is no central management agency. Again, many of the recommendations above apply to these fisheries. The following precautionary steps can also apply to some recreational fisheries:

a.   keep some areas closed to fishing in order to obtain the benefits noted above as item (d) under "New or Developing Fisheries". Also ensure that excessive fishing effort does not develop in the open areas;

b.   delegate some of the decision-making, especially area closures and entry limitations, to local communities or cooperatives;

c.   ensure that fishing pressure from other (e.g., industrial) segments of the fishery does not deplete the resources to the point where severe corrective action is needed, and

d.   investigate the factors that influence the behaviour of harvesters to develop approaches that can control fishing intensity. For example, improving incomes of individual harvesters may reduce pressure on resources.

4. PRECAUTIONARY APPROACH TO FISHERY RESEARCH

51.   Application of the precautionary approach to fishery management depends on the amount, type and reliability of information about the fishery and how this information is used to achieve management objectives. The precautionary approach to fishery management is applicable even with very limited information. Research to increase information about a fishery usually increases potential benefits while reducing the risk to the resource. The scientific and research input that is required for the precautionary

approach to fisheries is considered under the following headings; management objectives, observations and information base, stock assessment and analysis and decision processes.

4.1 The Role of Research in Establishing Management Objectives

52. There is a valid scientific role in helping managers develop objectives, so that scientific input to the overall management process is as effective as possible in achieving management intent. The precautionary approach requires continuing and anticipatory evaluation of the consequences of management actions with respect to management objectives. Scientific evaluation of consequences with respect to management objectives requires explicit definition of quantifiable criteria for judgement. An important scientific contribution is in the development of operational targets, constraints and criteria that are both scientifically usable and have management relevance.

53. Research is required to help formulate biological objectives, targets and constraints regarding the protection of habitat, the avoidance of fishing that significantly reduces population reproductive capacity, and reduces the effects of fishing on other (e.g., non-target) species. Combined with biological research, research on socio-economics and the structure of fishing communities is needed to formulate management objectives.

54. Until stock specific research leads to the establishment of alternative operational target based on research and practical experiences, a precautionary approach would seek to: (a) maintain the spawning biomass at a prudent level (i.e., above 50% of its unexploited level), (b) keep the fishing mortality rate relatively low (i.e., below the natural mortality rate), (c) avoid intensive fishing on immature fish, (d) protect the habitat.

4.2 Observation Processes and Information Base

55. A precautionary approach to fisheries requires explicit specification of the information needed to achieve the management objectives, taking account of the management structure, as well as of the processes required to ensure that these needs are met. Periodic evaluation and revision of the data collection system is necessary.

56. A precautionary approach would include mechanisms to ensure that, at a minimum, discarded catch, retained catch and fishing effort data are accurate and complete. These mechanisms could include use of observers and identification of incentives for industry co-operation.

57. Recognizing that resource users have substantial knowledge of fisheries, a precautionary approach makes use of their experience in developing an understanding of the fishery and its impacts.

58. The precautionary approach is made more effective by development of an understanding of the sources of uncertainty in the data sampling processes, and collection of sufficient information to quantify this uncertainty. If such information is available it can be

explicitly used in the management procedure to estimate the uncertainty affecting decisions and the resulting risk. If such information is not available, a precautionary approach to fishery management would implicitly account for the unknown uncertainty by being more conservative.

59. Precautionary fishery monitoring is part of precautionary research. It includes collection of information to address issues and questions that are not only of immediate concern but which may reasonably be expected to be important for future generations on in case objectives are changed. Information should be collected on target species, bycatch, harvesting capacity, behaviour of the fishery sector, social and economic aspects of the fishery, and ecosystem structure and function. Measures of resource status independent of fishery data are also highly desirable.

60. The precautionary approach relies on the use of a history of experience with the effects of fishing, in the fishery under consideration and/or similar fisheries, from which possible consequences of fishing can be identified and used to guide future precautionary management. This requires that both data and data collection methods are well documented and available.

61. There are many management processes and decision structures used throughout the world, such as regional management bodies, co-management, community-based management, and traditional management practices. Research is needed to determine the extent to which different management processes and decision structures promote precaution.

4.3 Assessment Methods and Analysis

62. Biological reference points for overfishing should be included as part of a precautionary approach.

63. A precautionary approach specifically requires a more comprehensive treatment of uncertainty than is the current norm in fishery assessment. This requires recognition of gaps in knowledge, and the explicit identification of the range of interpretations that is reasonable given the present information.

64. The use of complementary sources of fishery information should be facilitated by active compilation and scientific analysis of the relevant traditional knowledge. This should be accompanied by the development of methods by which this information can be used to develop management advice.

65. Specifically the assessment process should include:

   a.   scientific standards of evidence (objective, verifiable and potentially replicable), should be applied in the evaluation of information used in analysis;

   b.   a process for assessment and analysis that is transparent, and

   c.   periodic, independent, objective and in-depth peer review as a quality assurance.

66. A precautionary approach to assessment and analysis requires a realistic appraisal of the range of outcomes possible under fishing and the probabilities of these outcomes under different management actions. The precautionary approach to assessment would follow a process of identifying alternative possible hypotheses or states of nature, based on the information available, and examining the consequences of proposed management actions under each of these alternative hypotheses. This process would be the same in data-rich and data-poor analyses. A precautionary assessment would, at the very least, aim to consider: (a) uncertainties in data; (b) specific alternative hypotheses about underlying biological, economic and social processes, and (c) calculation of the theoretical response of the system to a range of alternative management actions. A checklist of issues for consideration under these headings is found in the following paragraphs.

67. Sources of uncertainty in data include: (a) estimates of abundance; (b) model structure; (c) parameter values used in models; (d) future environmental conditions; (e) effectiveness of implementation of management measures; (f) future economic and social conditions; (g) future management objectives, and (h) fleet capacity and behaviour.

68. Specific alternative hypotheses about underlying biological, economic and social processes to be considered include: (a) depensatory recruitment or other dynamics giving rapid collapse; (b) changes in behaviour of the fishing industry under regulation, including changes in coastal community structure; (c) medium-term changes in environmental conditions; (d) systematic underreporting of catch data; (e) fishery-dependent estimates of abundance not being proportional to abundance; (f) changes in price or cost to the fishing industry; and (g) changes in ecosystems caused by fishing.

69. In calculating (simulating) the response of the system to a range of alternative management actions, the following should be taken into account:

   a.   short-term (1-2y) projections alone are not sufficient for precautionary assessment; time frames and discount rates appropriate to inter-generational issues should be used, and

   b.   scientific evaluation of management options requires specification of operational targets, constraints and decision rules. If these are not adequately specified by managers, then precautionary analysis requires that assumptions be made about these specifications, and that the additional uncertainty resulting from these assumptions be calculated. Managers should be advised that additional specification of targets, constraints and decision rules are needed to reduce this uncertainty.

70. Methods of analysis and presentation will differ with circumstances, but effective treatment of uncertainty and communication of the results are necessary in a precautionary assessment. Some approaches (see also the Appendix to this section) that could prove useful are:

a. where there are no sufficient observations to assign probabilities to different states of nature that have occurred, decision tables could be used to represent different degrees of management caution through the Maximin and Minimax criteria;

b. where the number of different states of nature and the number of potential management actions considered are small, but probabilities can be assigned, decision tables can be used to show the consequences and probabilities of all combinations of these, and

c. where the range of states of nature is large, the evaluation of management procedures is more complex, requiring integration across the various sources of uncertainty.

71. A precautionary approach to analysis would examine the ability of the data collection system to detect undesirable trends. When the ability to detect trends is low, management should be cautious.

72. Since concern regarding the reversibility of the adverse impacts of fishing is a major reason for a precautionary approach, research on reversibility in ecosystems should be an important part of developing precautionary approaches.

4.4 Implementation Guidelines

73. The following measures could be applied in order to implement a precautionary approach to fishery research:

a. take into account the best scientific evidence available when designing and adopting management and conservation measures, in accordance with the provisions of the 1982 UNCLOS Convention. In the context of precautionary management, the best scientific evidence is described in section 4.3;

b. require a minimum level of information to be made available for any fishery to start or continue;

c. ensure that the *"lack of full scientific certainty shall be not used as a reason for postponing cost-effective measures to prevent environmental degradation"* (principle 15 of the Rio Declaration);

d. reduce critical uncertainties in the management plan;

e. take measures aimed at eliminating or reducing non-reporting and misreporting, *inter alia,* by ensuring that the fisheries sector cooperates in data collection and the public is fully informed of the results and uncertainty in the assessment;

f. systematically analyze various possible management options using the whole range of available models (bioeconomic, multispecies and behavioural), showing: (a) the likely direction and magnitude of the biological, social and economic consequences; (b) the related levels of uncertainty and the potential costs of the proposed action (risk assessment), and of no action (*status quo* scenarios);

g. promote multidisciplinary research, including: (a) social, economic and environmental sciences, and (b) research on management institutions and decision-making processes;

h. develop scientific information on multispecies and ecosystem processes as a foundation for identifying acceptable degrees of disturbance;

i. identify biological limit and target reference points for affected species and stocks, habitats and the ecosystem at large;

j. identify bioeconomic reference points to address the objectives of the fishery management plan;

k. improve methods for quantification of direct and indirect impacts of fishing;

l. improve understanding of the performance of different management structures in relation to precaution;

m. develop methods for optimizing the monitoring system, and

n. develop research and development programmes aiming at improving the performance of fishery technology in relation to environmental impacts and precautionary management.

Appendix

74. **The Minimax/Maximin approach** is a way of examining uncertainty and guiding decisions without the need for explicit statements of probabilities on the alternative hypotheses (Schmid, A. 1989, Cost-benefit analysis, West View Press). In the table below, S1 and S2 represent the alternative hypotheses about the resource (sometimes called "the different states of nature"). In this example, S1 is a hypothesis that implies a higher level of resource productivity and sustainable yield than S2. In the table rows, D1 to D3 represent alternative decisions. In this example D1, D2 and D3 broadly represent respectively a high, medium, and low level of fishing effort. The Ps represent the probabilities being placed on hypotheses being true. Values in the table represent the relative value of the outcome of a decision as applied to a given state of nature. In the example, these values could be regarded as representing sustainable catch.

Decision	S1 p = ?	S2 p = ?
D1	100	5
D2	50	40
D3	70	20

75. **The Maximin Values Criterion** is a cautious approach that leads to selects the maximum (highest) of the minimum outcome. The following table gives the relative value of outcomes for decisions given the hypothesis being true. Decision 2 would be chosen by this approach.

Decision	S1 p = ?	S2 p = ?	Minimum Value
D1	100	5	5
D2	50	40	40
D3	70	20	20

76. **The Minimax Regret Criterion** is a less cautious approach that selects the minimum of the maximum regret. The following table give a measure of regret for each decision when the hypothesis is true. Decision 3 would be chosen by this approach.

Decision	S1 p = ?	S2 p = ?	Minimum Regret
D1	100-100 = 0	40-5 = 35	35
D2	100-50 = 50	40-40 = 0	50
D3	100-70 = 30	40-20 = 20	30

77. **The decision table approach** uses the probabilities of alternative hypotheses of the state of nature, along with the values of the outcomes of decisions to give an expected value and variance of each action across all alternative hypotheses.

Decision	S1 p = 0.7	S2 p = 0.3	Expected Value	Variance
D1	100	5	71.5	1895.25
D2	50	40	47.0	21
D3	70	20	55.0	525

78. The decision is to select the desired trade-off between the variance and the expected value of the outcome. The variance is a measure of uncertainty and the expected value

gives the expected result of choosing a given policy. Policy D1 has a high expected outcome with the highest uncertainty, and a 0.3 probability of a very poor outcome. On the other hand, policy D2 has a much lower expected outcome with lower uncertainty, and no probability of a poor outcome.

79.   Methods that integrate across uncertainty: Where the number of possible states of nature is large, as is almost always the case, a mathematical equivalent operation to that described above for a simple decision table can be carried out. This results in the calculation of outcome probabilities for each possible decision, integrating across uncertainties.

5. PRECAUTIONARY APPROACH TO FISHERY TECHNOLOGY

5.1 Objective

80.   Recognizing that many aquatic resources are overfished and that the fishing capacity presently available jeopardize their conservation and rational use, technological changes aimed solely at further increasing fishing capacity would not generally be seen as desirable. Instead a precautionary approach to technological changes would aim at:

a.   improving the conservation and long-term sustainability of living aquatic resources;

b.   preventing irreversible or unacceptable damage to the environment;

c.   improving the social and economic benefits derived from fishing, and

d.   improving the safety and working conditions of fishery workers.

5.2 Introduction

81.   Fishery technology consists of the equipment and practices used for finding, harvesting, handling, processing and distributing of aquatic resources and their products.

82.   Different fishery technologies will have different effects on the ecosystem, the social structure of fishing communities, the safety of fishery workers and the ease, effectiveness and efficiency of management of the fishery. It is the amount and context in which fishery technology is used (e.g. when, where and by whom) that influence whether the objectives of fisheries management are reached, and not the technology. For instance, the current overfishing of many aquatic resources is the product of both the efficiency of the finding and catching technologies and of the amount used. Similarly, building a fishmeal plant might involuntarily result in severe changes in the way the fishery is conducted, and in the community's social structure.

83. Fishery technology is constantly evolving and its efficiency in catching fish will increase over time. For example, a 4% increase in efficiency per year would cause a doubling of the fishing mortality rate in 18 years if the fishing effort remained constant. A precautionary approach to management should take such increases into account.

84. A precautionary approach should be adopted for the development of new technologies or the transfer of existing technologies to other fisheries to avoid unplanned abrupt changes in fishing pressure or social structures. Certain technologies will be considered undesirable, if they create unacceptable effects (e.g., poison or explosives) or if their adoption leads to wasteful use (e.g., at sea, sorting machines have been banned where they might increase discarding).

85. Fishery technologies produce side effects on the environment and on non-target species. These effects have often been ignored but, in the context of a precautionary approach, some technologies may warrant a review. Similarly, a precautionary approach would encourage careful consideration of the side effects of new fishery technologies before they are introduced.

86. Each fishery technology has advantages and disadvantages that should be balanced in a precautionary approach, and it may be better to have a mixture of technologies. When new fishery technology is introduced, it should be carefully evaluated to assess its potential direct and indirect effects. If a mix of fishery technology representing "best current practice" in an area can be identified, precautionary management would encourage its adoption while it would discourage damaging ones. Responsible fishery technology achieves the specific fishery management objectives with the minimumdamaging side effects. These concepts (of responsible fishing and best current practices) were addressed by the UN General Assembly[1] and in the Cancun Declaration[2].

87. A precautionary approach would provide for a process of initial and on-going review of the effects of fishery technology as it is introduced or evolves in local practice. However, the extent to which a precautionary approach can be applied to the management of technological changes depends on the existing level of management. In some cases, education of fishermen and consumers towards responsible practices may be the only possible approach. Where elaborate research, management and enforcement systems are in place, a wider variety of options are available for application of the precautionary approach. However, although some gears and

---

[1]General Assembly resolution 44/228 of 22 December 1989 on UNCED referred instead to "*environmentally sound technology*", stressing the need for socio-economic constraints to be taken into account. The wording does not pretend to limit the choice to a single "best" or soundest technology, implying that many "sound" technologies may be used together, depending on the socio-economic context of their introduction

[2] The Cancun Declaration (Mexico, 1992) provides that "*States should promote the development and use of selective fishing gear and practices that minimise waste of catch of target species and minimise by-catch of non-target species*", focusing on only one aspect of responsible fishing technology

practices are prohibited they may continue to be used. The adoption of a precautionary approach to the management of new fishery technology depends on the ability to achieve compliance through education and/or enforcement. The following sections assume that institutional arrangements exist to achieve compliance.

5.3 Evaluating the Impacts of Technologies

88.   A precautionary approach to developing and selecting responsible technologies for fishing requires an appropriate understanding of the consequences of their adoption and use. These consequences, particularly the impacts on non-target species and ecosystems, may be highly uncertain. Nevertheless, some information exists and more can be obtained. The problem of evaluating impacts is relevant both to the use of existing technologies and to the development of new ones, as well as to the introduction of existing technologies to new areas. The description of a given technology would state its relative impacts and advantages for a given species in a specific environment. Target fishery, environmental and ecosystem, socio-economic and legal factors should be considered when evaluating the impacts of fishery technologies.

89.   The factors to consider when evaluating the impacts of fishery technology include:

a.   target-fishery factors such as selectivity by size and species (e.g., target, non-target, and protected species; discards; survival of escapees; "Ghost fishing"; and catching capacity;

b.   environmental and ecosystem factors such as bio-diversity; habitat degradation; contamination and pollution; generation of debris and rubbish disposal; direct mortality; predator-prey relationships;

c.   socio-economic factors such as safety and occupational hazards; training requirements; user conflicts; economic performance; employment; monitoring and enforcement requirements and costs; and techno-economic factors (i.e., infrastructure and service requirements; cost and technological accessibility; product quality; and energy efficiency), and

d.   legal factors such as existing legislation; need for new legislation; international agreements; and civil liberties.

90.   These factors could be used to identify beneficial new technologies or damaging ones, to assess the ability of a fishery to accommodate increased use of an established technology and to help direct monitoring and special reporting procedures towards important questions. Technologies for aids to navigation, fish-locating devices, processing and distribution could also be described and evaluated using the above criteria. This will require a suitable description of technologies, cross-referenced against a range of possible impacts. Other elements relevant to the specific technology/area evaluated would also be included.

91.   The approaches used to evaluate impacts will vary according to the human and financial resources available to collect information. If resources are limited, it may be possible

to make decisions based on existing information on the impacts of similar technologies in similar environments. Monitoring of existing fishing practices (for example recording of bycatch) will provide additional information.

92.   Where financial and human resources are limited, existing information on impacts could be used to do desk studies following the approach to evaluation suggested above. Although some general guidelines can be given, based on known characteristics of types of resources and technology, the most appropriate mix of technologies to be used in a particular fishery should be established on a case-by-case basis, following evaluations made at appropriate regional and national levels. Such evaluations could be refined with practical experience and weighed in accordance with local social and economic values.

93.   In the case of new technologies, or technologies new to an area, pilot studies may be cost-effective in evaluating the impacts and can be useful in demonstrating the benefits of new technology. For example, the introduction of escape ports in lobster traps for undersized individuals demonstrated to fishermen that catch rates of large lobsters increased. On the other hand, pilot studies cannot demonstrate long-term gains such as increased yield per recruit, but they will show the short-term losses.

94.   Considerable resources are required for major experiments to measure effects of fishery technology on the marine environment, but well-designed experiments of this type (either as research projects or via experimental management) will provide the most useful information on which to judge the impacts of technologies in particular areas or habitats. This information may be relevant in other areas than the study sites or fisheries from which the data were derived.

95.   Procedures developed in other contexts for protecting the environment[3] could also be suitable when evaluating new technologies in fisheries or major alterations to existing ones. This would be particularly necessary when there are vulnerable resources or fragile ecosystems, that must be protected. In a precautionary approach, proponents of new fishery technology would be required by the State to provide for a proper evaluation of the potential impacts of new techniques before authorization is given.

96.   The maximum cost that could be justified for evaluating new fishery technology or practices should be in proportion to the expected benefits and impacts.

---

[3]Before introducing a possibly dangerous technology or discharging pollutants, industries have to provide information on the potential impact in order to obtain a permit from authorities. Usually a number of special measures are prescribed for monitoring the effect and limiting the possible impacts on the environment. A softer approach is the Prior Informed Consent (PIC), a more stringent one the Prior Consultation Procedures (PCP); the former mainly requires a consent from those who could possibly become affected, while the latter is a more formal procedure. Those mechanisms however are efficient only when there is a powerful and competent environmental authority

97.   In a precautionary approach to managing fishery technology, a designated lead authority should have the mandate to evaluate and decide on the acceptability of a proposed new technology, or changes to existing technology, and oversee the impact evaluation procedure. Proponents and other stakeholders should be able to appeal if the proper procedure has not been followed or if the decision by authorities does not appear to agree with the conclusions of the review.

98.   As authorization procedures in the majority of cases would be for minor technical improvements, the procedures could be kept simple and administration costs held at a relatively low level. However, minimal progressive improvements will accumulate over time and periodic reviews of the impacts of existing technology will be necessary. Increases in catching efficiency result from the rapid growth in the use of modern information technologies in most fisheries around the world (acoustic fish detection and identification, gear and vessel monitoring, satellite-based environmental sensing and navigation, and easy inter-vessel communication). However, information, formally treated as "a measure of the reduction of uncertainty", can also potentially improve selectivity, safety and profitability of fishing operations and thus create beneficial effects.

99.   Restricting the use of improved information technologies will rarely be justified or successful and there should be a positive attitude towards technical progress in fisheries in general especially with regards to safety at sea and fishermen's health.

100.   The benefits of technological improvements need adequate extension work and education to encourage their adoption. The promotion of the best technology would benefit from improvement in international cooperation regarding technology transfer, as underscored in UNCED's Agenda 21. The successful international efforts in the Eastern Central Pacific in training crews in effectively avoiding bycatches of dolphins through the use of specifically designed technology is a good example of what can be achieved in this respect.

5.5 Technology Research and Development

101.   Fishery technology research in support of a precautionary approach would encourage the improvement of existing technologies and promote the development of appropriate new technologies. Such research would not just concentrate on gears used for capture; for example, research into the cost-effective purification of water supplies to ice plants might considerably reduce post-harvest losses and improve product quality and safety.

102.   Technological developments such as satellite tracking may also help precautionary management by improving monitoring of commercial operations and by enabling research to reduce uncertainty about relevant aspects of fisheries science.

103.	The following measures could be applied in order to implement a precautionary approach to fishery technology development and transfer.

Authority

a.	Effective mechanisms to ensure that the introduction of technology is subject to review and regulation should be established.

Evaluation procedures

b.	A first step in the evaluation procedure is the documentation of the characteristics and amount of the fishery technology currently used.

c.	Procedures for the evaluation of new technologies with a view to identify their characteristics in order to promote the use of beneficial technologies and prevent usage of those leading to difficult-to-reverse changes should be established.

d.	These procedures should evaluate with appropriate accuracy the possible impacts of the proposed technology in order to avoid wasteful capital and social investments.

e.	Authorities should ensure that proponents and other stakeholders understand their obligations and their rights regarding such procedures.

f.	The extent of the evaluation procedures should match the potential effects of the proposed technology, e.g., from desk study through full scale impact studies, possibly including or leading to pilot projects.

Implementation

g.	Authorities should implement technology gradually to minimize the risk of irreversible damage or overinvestment.

h.	Existing technologies and their effect on the environment should be reviewed periodically.

i.	Technological developments may modify the practices of fishery workers. To achieve the full benefits of the technology and to ensure the safety of fishery workers, training in the proper use of the new technology should be provided.

j.	In fisheries that are being rehabilitated, the opportunity should be taken to review the mix of technologies used.

k.	Research into responsible fishery technology should be encouraged.

l.	Technology research for the reduction of uncertainty in stock assessment and monitoring should be encouraged.

6. PRECAUTIONARY APPROACH TO SPECIES INTRODUCTION

6.1 Introduction

104. Because of the high probability that impacts of species introduction be of irreversible and unpredictable impacts, many species introductions are not precautionary. Therefore, a strictly precautionary approach would not permit deliberate introductions and would take strong measures to prevent unintentional introductions. Recognizing the difficulties with introductions, the objectives of a precautionary approach to species introductions in relation to capture fisheries should be to reduce the risk of adverse impacts of introductions on capture fisheries, to establish corrective or mitigating procedures (as in a contingency plan) in advance of actual adverse effects, and to minimize unintended introductions to wild ecosystems and associated capture fisheries.

105. In relation to aquaculture, experience has shown that animals will usually escape the confines of a facility. As a consequence, the introduction of aquatic organisms for aquaculture should be considered as a purposeful introduction into the wild, even though the quarantine/hatchery facility may be a closed system.

106. Introductions and transfers (hereafter referred to as introductions) are an effective means to increase protein, generate income and provide employment. However, some intended and many unintended introductions may result in significant and serious impacts on capture fisheries. The numbers of unintended introductions, for example by means of ballast water, greatly outnumber those purposefully introduced for capture fisheries. In the case of introduced species for fishery purposes, the risk to capture fisheries can be reduced by the use of internationally accepted codes, such as the 1994 ICES Code of Practice (see Annex A)[4]. This code forms a basis for a more precautionary introduction and should be widely circulated and explained.

107. For a precautionary approach to fishery management, irreversible changes in the time scale of human generations and other undesirable impacts should be avoided, taking into account uncertainty. Species introductions, either purposeful or unintended, may have such undesirable effects. Once a species has been introduced, it cannot usually be eradicated, although it may be possible to mitigate its undesirable effects.

108. The difficulty in reversing an introduction and its adverse effects should figure prominently in the decision process on whether to allow an introduction. Assessment of the risks of intentional introductions on fisheries is necessary for the precautionary approach; the ICES Code of Practice provides a procedure for such an assessment.

---

[4]**Editor's note:** The ICES Code of Practice originally drafted by ICES was subsequently finalized jointly by ICES and EIFAC for use by the FAO Regional Fishery Bodies. The Code is still being evolved and supplementary material is being produced by FAO to assist in its implementation mainly in developing countries

109. To encourage more compliance with the precautionary approach to introductions, the existing Code of Practice formulated for the ICES region can be modified and adapted to suit more national implementation of the codes by streamlining its procedures without weakening the rigour of the codes.

110. Unintended introductions are inherently unprecautionary because they can rarely be evaluated in advance. A precautionary approach would aim at reducing the risk of such unintended introductions and minimize their impact.

6.2 Main Issues and Objectives

111. Introductions are considered here from the perspective of the fishery sector. The main reasons for deliberate introductions include production of protein, employment, generation of foreign exchange, biological control and recreation. Species have been introduced through activities associated with transport (i.e., ballast water, ship and oil-rig fouling), trade in living organisms including aquarium species, aquaculture and fisheries (commercial, recreational, stock enhancement, organisms carried on fishing equipment, live bait fish). Many of these activities have increased over the last century and are expected to increase further in the future.

112. Potential impacts of some introductions on the fishery sector included changes in the distribution and abundance of fishery resources through disease, changes in predator-prey relationships, changes in competition, mixing of bad (maladapted) genes, and habitat modification. There may also be second and third order changes that affect the ecosystem. Changes in fishing strategy and the fishing community may also require changes in fishing gear and season to allow for a newly introduced species to establish itself or to avoid potential side issues associated with the new fishery. Climate changes may also have significant consequences that may modify the environment, making it more suitable for introductions of either useful or harmful species.

113. The use of introduced species, including genetically modified and genetically selected organisms, may allow for continued or increased production from habitats that have been so altered or degraded that native fisheries are no longer viable. Care should be taken not to use this potential productivity from introduced species as justification for further abuse of habitat or for delaying their restoration.

6.3 Research and Technology

Deliberate introductions

114. The ICES Code of Practice (Annex A) describes the research activities that should be conducted in advance of an introduction as follows: (1) desk assessment of the biology and ecology of the intended introduction; (2) preparation of a hazard assessment (detailed analysis of potential environmental impacts); (3) examination of the species within its home range. The results of the above research should be contained in a prospectus or proposal to be submitted to the competent authority for evaluation and decision.

115. Technological intervention can be used during and following the introduction; such technologies may include:

   a.   use of hatcheries and quarantine stations to reduce the chance of spread of disease to fishery resources, and to impart some control on numbers of exotic organisms released;

   b.   use of sterile organisms to reduce the chance of interbreeding with natural fish stocks;

   c.   genetic stock identification to reduce or prevent genetic changes in the fishery resource;

   d.   disease diagnostics to monitor the health of the introduced species, and

   e.   development of the use of limited (pilot) scale introductions to assess impacts and performance.

116. Continued research and technological intervention on introduced species as part of a monitoring and evaluation scheme should be conducted to assess their impact, health and performance within their new habitat. In this regard, databases or registries of introductions of aquatic species, including their ecological and socio-economic impact, have been established and should be maintained by competent organizations with input from the fishery sector.

Unintended introductions

117. Although unintended introductions may arise from several sources, such as fouling organisms, removal of natural barriers and aquarium fish trade, ballast water is probably the most significant and troublesome for the fishery sector and, therefore, emphasized here. In the case of ballast water and sediment, desk studies may be undertaken to determine (1) main ballast water sources, (2) volumes of ballast introduced, and (3) likely "hot spots" as sources of introductions.

118. Active research should take place and continue on:

   a.   practical methods for treating organisms in ballast water and sediment;

   b.   study of dynamics of target species in voyage;

   c.   study of algal cysts in ships ballast sediment and in port areas;

   d.   effectiveness of reballasting activities;

   e.   design changes to ballast water tanks to kill or control harmful species in ballast water, and

f. vessel design that will facilitate the treatment and handling of ballast sediments and water.

119. Research into new effective non-biocidal antifouling materials should continue to reduce the risks of introductions on ships' hulls and to replace those biocidal applications detrimental for capture fisheries. Antifouling agents are designed to reduce drag and increase the fuel efficiency of a vessel and its long-term efficacy, but they should also be environmentally friendly. Consideration should also be given to those antifouling agents designed to control organisms that would be especially harmful to capture fisheries, even though they may not affect the performance of the vessel.

6.4 Management

120. The first step to managing introductions is to establish a management authority with responsibility for evaluation of proposed introductions, approval in accordance with these guidelines and assuring monitoring of the effects of the introduction. The management of species introductions will involve comparative risk assessment and choices between various options to increase productivity. Management options are limited here to those in the aquatic sector, although countries may be aware of broader issues, such as the development of other sectors (e.g., agriculture). International codes of practice, such as the ICES code, provide a good framework for the management of purposeful species introductions. These codes should not be seen as a hindrance to development, but rather as a tool to help importers make good choices with regard to introductions. Implementation of the code should increase the probability of success of an appropriate introduction.

Deliberate introductions

121. Intended introductions should be controlled. As a consequence, those making an introduction should follow the ICES or similar code of practice as appropriate and would be expected to demonstrate caution by preparing a proposal covering: (1) the purpose and objectives of the introduction in advance of the introduction, (2) all relevant biological, ecological, and genetic data of the species in the target area likely to be affected, (3) analysis of potential impacts at the introduction site, including potential ecological, genetic and disease impacts and consequences of its spread, and (4) a qualitative and, where possible, a quantitative risk assessment.

122. If this proposal is approved: (1) a brood stock should be established at a suitable quarantine site; (2) all effluents from facility should be appropriately sterilized; (3) isolated first generation individuals, free of disease, should be released to the wild in small numbers; and (4) studies of the introduction in the new environment should be continued.

123. A contingency plan should take account of negative effects should they become apparent and warrant intervention.

124. The code should also cover introductions that are part of current commercial practice (live trade in fish and shellfish) and recommend: (a) periodic inspection prior to exportation; (b) regular inspection; and (c) quarantine and control if appropriate.

125. The concern expressed in relation to the introduction of species for fisheries, using the ICES Code, should apply to those species under consideration for biological control, which may have implications for capture fisheries. Bio-control programmes should be weighed carefully against other control methodologies, such as physical and chemical techniques or through intensive fishing. It is likely biological control techniques will take some years to evaluate, through field trials. Much may be learned from the studies on biological control in other disciplines, such as entomology.

126. The proposal submitted by the potential importer of an exotic species and its review by the competent authority serve as precautionary measures to reduce the chance of a harmful introduction. Governments may wish to consider for national legislation that, if these elements of the code are not followed, the importer of an exotic species may be financially responsible and subject to liability, should significant negative effects arise.

127. Care should be taken to ensure that the introduced population has an adequate genetic resource base, i.e., genetic diversity, low inbreeding, etc. This may reduce the need for additional introductions, which might otherwise be necessary to increase the genetic resource base. In addition, consideration should be given to the use of gametes, e.g., eggs, cryopreserved sperm, as import material instead of whole organisms to reduce the risk of introducing disease or unintended organisms.

Unintended introductions

128. Introduction of undesirable species via ballast water poses problems for fisheries worldwide. In addition to unintentional introductions by ballast water, there are many other mechanisms, including fishing and trade in live fish. Fishing can introduce species by transporting live or fresh bait, or biologically contaminated fishing gear between ecosystems. With trade in live organisms (for aquarium or human consumption), there is the risk of escape.

129. Authorities responsible for regulating fishing and trade in fish products should establish regulations to reduce these risks, commensurate with the severity of potential adverse impacts. However, the national and international competent authorities that deal with ballast issues are seldom also responsible for fishery management matters. Cooperation between these authorities would greatly aid the management of this problem.

130. In order to reduce the risk of introductions of organisms in ballast water on capture fisheries in or near deballasting areas, the following methods of prevention include, as recommended by the IMO (1994, Annex B): (a) non-release of ballast; (b) ballast water exchange(s) in or near approved areas; (c) preventing or minimizing uptake of contaminated water or sediment (in shallow water, near dredging operations, during

algal blooms); (d) special ballasting facilities on shore; (e) education of crews about ballast-water management procedures, and (f) treatments of ballast water, including changes in temperature and salinity and use of biocides (chemicals).

131. Although the issues of ballast-water transport, fouling organisms and other unintentional introductions may fall outside the mandate of fishery ministries, the fishery sector could contribute to the management of such introductions, which impact upon the industry. This could be accomplished by promoting the establishment and maintenance in the appropriate institution, of an accessible database on ballast or fouling organisms that have a demonstrated impact on fisheries, by promoting a network of experts who would identify problems, assist with species identification, and delimit areas of impact. The fishery sector may be well placed to detect the spread of harmful ballast/fouling organisms and should, therefore, contribute to such databases and networks once established, and may take a lead in instigating action on environmental management.

132. Introduced organisms may cause major changes in ecosystems, especially in port and associated, partly enclosed or enclosed areas, such as lagoons. Such introductions may result in changes in the productivity of local harvested species. Monitoring of introduced organisms and fisheries in these areas may provide useful information as a basis for modifying management techniques and policy for harvested resources.

6.5 Implementation Guidelines

133. Those making introductions should consider the ICES code as a means to reduce introductions of harmful or nuisance species, including parasites and diseases, which may impact on capture fisheries:

134. To encourage a more precautionary approach, governments outside the ICES region should follow the principles or recommendations of the code according to their particular circumstances. Critical elements to be considered are a proposal, independent review by a scientific body, and subsequent protocols if an introduction is approved.

135. In addition to codes of practice issued by ICES and IMO, the following may be useful as a precautionary approach to introduced species:

  a. establish clear and straightforward procedures and protocols on the mechanisms for the management of introduced species under the relevant competent government agencies with authority to address issues of compliance, responsibility, and liability;

  b. promote cooperation between the fishery sector and other sectors dealing with the aquatic environment in order to coordinate policy and regulation of introduced species, especially the national shipping sector, and port authorities and international organizations, e.g., ICES, FAO, IMO and ICLARM, which have relevant expertise;

c. encourage relevant groups (e.g., importer and regulatory agencies) to consider the development of a contingency plan in the event that the introduced species does not fulfil expectations or causes adverse impacts;

d. promote education, training and awareness of harmful species introductions, disseminate as widely as possible the ICES and IMO codes of practice and advise responsible authorities within the fishery and other sectors on the procedures of these codes;

e. develop an international information system in appropriate institutions on ballast or fouling organisms, which have demonstrated impacts on fisheries, by promoting a network of experts who would identify problems, species identification, and areas of impact, arrange standardization of sampling methods (inter-calibration), and develop a monitoring system so that changes can be evaluated in high-risk areas. Should unwanted species be detected, an eradication programme should be considered;

f. encourage compliance with the code by the fishing industry and other users of aquatic resources; national governments could encourage self-policing and self-enforcing by fishing industry and other users of the resource in minimising the impacts of introducing species by unauthorised means, and

g. conduct research on the applicability of information gained from introductions of limited numbers of animals (e.g. pilot/experimental introductions).

136. Promotion and maintenance of databases on species deliberately introduced for capture fisheries is suggested. This would include the impact of these introductions. Importers or fishery managers may wish to consult such databases to assist in the proposal formulation and its evaluation.

137. The development of effective non-biocidal antifouling applications to reduce the risk of introduction from ship fouling is encouraged.

CODE OF PRACTICE ON THE INTRODUCTION AND TRANSFER
OF MARINE ORGANISMS, 1994[5]
(ICES, 1995)

The introduction and transfer of marine organisms, including genetically modified organisms, carry the risk of introducing not only pests and disease agents but also many other species. Both intentional and unintentional introductions may have undesirable ecological and genetic effects in the receiving ecosystem, as well as potential economic impacts. This Code of Practice provides recommendations for dealing with new intentional introductions, and also recommends procedures for species which are part of existing commercial practice, in order to reduce the risks of adverse effects that could arise from such movements.

I   **Recommended procedure for all species prior to reaching a decision regarding new introductions.** (A recommended procedure for introduced or transferred species which are part of current commercial practice is given in Section IV; a recommended procedure for the consideration of the release of genetically modified organisms is given in Section V.)

    (a)   Member Countries contemplating any new introduction should be requested to present to the Council at an early stage a detailed prospectus on the proposed new introduction(s) for evaluation and comment.

    (b)   The prospectus should include the purpose and objectives of the introduction, the stage(s) in the life cycle proposed for introduction, the area of origin and the target area(s) of release, and a review of the biology and ecology of the species as these pertain to the introduction (such as the physical, chemical, and biological requirements for reproduction and growth, and natural and human-mediated dispersal mechanisms).

    (c)   The prospectus should also include a detailed analysis of the potential impacts on the aquatic ecosystem of the proposed introduction. This analysis should include a thorough review of:

         i) the ecological, genetic, and disease impacts and relationships of the proposed introduction in its natural range and environment;

         ii) the potential ecological, genetic, and disease impacts and relationships of the proposed introduction in the proposed release site and environment. These aspects should include but not necessarily be limited to:

---

[5]Reproduced for easy reference by courtesy of the International Council for the Exploration of the Sea (ICES)

- potential habitat breadth,
- prey (including the potential for altered diets and feeding strategies),
- predators,
- competitors,
- hybridization potential and changes in any other genetic attributes, and
- the role played by disease agents and associated organisms and epibiota.

Potential predation upon, competition with, disturbance of, and genetic impacts upon, native and previously introduced species should receive the utmost attention. The potential for the proposed introduction and associated disease agents and other organisms to spread beyond the release site and interact with species in other regions should be addressed. The effects of any previous intentional or accidental introductions of the same or similar species in other regions should be carefully evaluated.

(d) The prospectus should conclude with an overall assessment of the issues, problems, and benefits associated with the proposed introduction. Quantitative risk assessments, as far as reasonably practicable, could be included.

(e) The Council should then consider the possible outcome of the proposed introduction, and offer advice on the acceptability of the choice.

II **If the decision is taken to proceed with the introduction, the following action is recommended:**

(a) A brood stock should be established in a quarantine situation approved by the country of receipt, in sufficient time to allow adequate evaluation of the stock's health status.

The first generation progeny of the introduced species can be transplanted to the natural environment if no disease agents or parasites become evident in the first generation progeny, but not the original import. In the case of fish, brood stock should be developed from stocks imported as eggs or juveniles, to allow sufficient time for observation in quarantine.

(b) The first generation progeny should be placed on a limited scale into open waters to assess ecological interactions with native species.

(c) All effluents from hatcheries or establishments used for quarantine purposes in recipient countries should be sterilized in an approved manner (which should include the killing of all living organisms present in the effluents).

(d) A continuing study should be made of the introduced species in its new environment, and progress reports submitted to the International Council for the Exploration of the Sea.

III   Regulatory agencies of all Member Countries are encouraged to use the strongest possible measures to prevent unauthorized or unapproved introductions.

IV   Recommended procedure for introduced or transferred species which are part of current commercial practice.

    (a)   Periodic inspection (including microscopic examination) of material prior to exportation to confirm freedom from introducible pests and disease agents. If inspection reveals any undesirable development, importation must be immediately discontinued. Findings and remedial actions should be reported to the International Council for the Exploration of the Sea.

and/or

    (b)   Quarantining, inspection, and control, whenever possible and where appropriate.

    (c)   Consider and/or monitor the genetic impact that introductions or transfers have on indigenous species, in order to reduce or prevent detrimental changes to genetic diversity.

It is appreciated that countries will have different requirements toward the selection of the place of inspection and control of the consignment, either in the country of origin or in the country of receipt.

V   Recommended procedure for the consideration of the release of genetically modified organisms (GMOs).

    (a)   Recognizing that little information exists on the genetic, ecological, and other effects of the release of genetically modified organisms into the natural environment (where such releases may result in the mixing of altered and wild populations of the same species, and in changes to the environment), the Council urges Member Countries to establish strong legal measures[6] to regulate such releases, including the mandatory licensing of physical or juridical persons engaged in genetically modifying, or in importing, using, or releasing any genetically modified organism.

    (b)   Member Countries contemplating any release of genetically modified organisms into open marine and fresh water environments are requested at an early stage to notify the Council before such releases are made. This notification should include a risk assessment of the effects of this release on the environment and on natural populations.

---

[6]Such as the European Community Council Directive of 23 April 1990 on the deliberate release into the environment of Genetically Modified Organisms (90/220/EEC) Official Journal of the European Communities, L117: 15-17 (1990)

(c)   It is recommended that, whenever feasible, initial releases of GMOs be reproductively sterile in order to minimize impacts on the genetic structure of natural populations.

(d)   Research should be undertaken to evaluate the ecological effects of the release of GMOs.

DEFINITIONS

For the application of this Code, the following definitions should be used:

Brood stock

Specimens of a species, either as eggs, juveniles, or adults, from which a first or subsequent generation may be produced for possible introduction to the environment.

Country of origin

The country where the species is native.

Current commercial practice

Established and ongoing cultivation, rearing, or placement of an introduced or transferred species in the environment for economic or recreational purposes, which has been ongoing for a number of Years.

Disease agent

For the purpose of the Code, 'disease agent' is understood to mean all organisms, including parasites, that cause disease. (A list of prescribed disease agents, parasites, and other harmful agents is made for each introduced or transferred species in order that adequate methods for inspection are available. The discovery of other agents, etc., during such inspection should always be recorded and reported.)

Genetic diversity

All of the genetic variation in an individual, population, or species. (ICES, 1988)

Genetically modified organism

(GMO) An organism in which the genetic material has been altered anthropogenically by means of gene or cell technologies[7].

---

[7]Such technologies include the isolation, characterization, and modification of genes and their introduction into living cells or viruses of DNA as well as techniques for the production of living cells with new combinations of genetic material by the fusion of two or more cells

Introduced species (= non-indigenous species, = exotic species)

Any species intentionally or accidentally transported and released by humans into an environment outside its present range.

Marine species

Any aquatic species that does not spend its entire life cycle in fresh water.

Quarantined species

Any species held in a confined or enclosed system that is designed to prevent any possibility of the release of the species, or any of its disease agents or any other associated organisms into the environment

Transferred species (= transplanted species)

Any species intentionally or accidentally transported and released within its present range.

.NOTES

(a)   It is understood that an introduced species is what is also referred to as an introduction, and a transferred species as a transfer.

(b)   Introduced species are understood to include exotic species, while transferred species include exotic individuals or populations of a species.

(c)   It is understood for the purpose of the Code that introduced and transferred species may have the same potential to carry and transmit disease or any other associated organisms into a new locality where the disease or associated organism does not at present occur.

REFERENCES

ICES. 1984. Guidelines for Implementing the ICES Code of Practice Concerning Introductions and Transfers of Marine Species. Cooperative Research Report No. 130. 20 pp.

ICES. 1988. Codes of Practice and Manual of Procedures for Consideration of Introductions and Transfers of Marine and Freshwater Organisms. Cooperative Research Report No. 159. 44 pp.

ICES. 1994. Report of the ICES Advisory Committee on the Marine Environment, 1994, Annex 3. ICES Cooperative Research Report No. 204. 122 pp.

GUIDELINES FOR PREVENTING THE INTRODUCTION OF UNWANTED AQUATIC ORGANISMS AND PATHOGENS FROM SHIPS' BALLAST WATER AND SEDIMENT DISCHARGES (IMO, 1994)[8]

1. INTRODUCTION

1.1   Studies carried out in several countries have shown that many species of bacteria, plants, and animals can survive in a viable form in the ballast water and sediment carried in ships, even after journeys of several weeks' duration. Subsequent discharge of contaminated ballast water or sediment into the waters of port States may result in the establishment of unwanted species which can seriously upset the existing ecological balance. Although other media have been identified for transferring organisms between geographically separated water bodies, ballast water discharge from ships appears to have been among the most prominent. The introduction of diseases may also arise as a result of port State waters being inoculated with large quantities of ballast water containing viruses or bacteria, thereby posing health threats to indigenous human, animal and plant life.

1.2   The potential for ballast water discharges to cause harm, was recognized by resolution 18 of the International Conference on Marine Pollution, 1973, from which Conference emerged the MARPOL Convention. Resolution 18 called upon the World Health Organization, in collaboration with the International Maritime Organization, to carry out research into the role of ballast water as a medium for the spreading of epidemic disease bacteria.

1.3   It is the aim of these Guidelines to provide Administrations and port State Authorities with guidance on procedures that will minimize the risk from the introduction of unwanted aquatic organisms and pathogens from ships' ballast water and sediment. The selection of an appropriate procedure will depend upon several factors, including the type or types of organisms being targeted, the level of risks involved, its environmental acceptability, and the economic and ecological costs involved.

1.4   The choice of procedures will also depend upon whether the measure is a short-term response to an identified problem or a long-term strategy aimed at completely eliminating the possibility of the introduction of species by ballast water. In the short term, operational measures such as ballast water exchange at sea may be appropriate where they have been shown to be effective and are accepted by port State Authorities and Administrations. For the longer term, more effective strategies, possibly involving structural or equipment modifications to ships, may need to be considered.

---

[8]Resolution A.774(18) Adopted on 4 November 1993. Reproduced by courtesy of the International Maritime Organization (IMO)

2. DEFINITIONS

For the purposes or these Guidelines, the following definitions apply:

Administration means the Government of the State under whose authority the ship is operating.

Member States means States that are Members of the International Maritime Organization.

Organization means the International Maritime Organization (IMO).

Port State authority means any official or organization authorized by the Government of a port State to administer guidelines or enforce standards and regulations relevant to the implementation of national and international shipping control measures.

3. APPLICATION

The Guidelines can apply to all ships; however, a port State Authority shall determine the extent to which these Guidelines do apply.

4. GENERAL PRINCIPLES

4.1   Member States may adopt ballast water and sediment discharge procedures to protect the health of their citizens from foreign infectious agents, to safeguard fisheries and aquaculture production against similar exotic risks and to protect the environment generally.

4.2   Application of ballast water and sediment discharge procedures to minimize the risk of importing unwanted aquatic organisms and pathogens may range from regulations based upon quarantine laws to guidelines providing suggested measures for controlling or reducing the problem.

4.3   In all cases, a port State Authority must consider the overall effect of ballast water and sediment discharge procedures on the safety of ships and those on board. Regulations or guidelines will be ineffective if compliance is dependent upon the acceptance of operational measures that put a ship or its crew at risk.

4.4   Ballast water and sediment discharge procedures should be practicable, effective, designed to minimize cost and delays to ships, and based upon these Guidelines whenever practicable.

4.5   The ability of aquatic organisms and pathogens to survive after transportation in ballast water may be reduced if significant differences in ambient conditions prevail - e.g. salinity, temperature, nutrients and light intensity.

4.6   If fresh water (FW), brackish water (BW) and fully saline water (SW) are considered, the following matrix provides, in most cases, an indication of the probability that aquatic organisms and pathogens will survive after being transferred.

Probability of organism's survival and reproduction

Receiving Waters	Discharged Ballast		
	FW	BW	SW
FW	HIGH	MEDIUM	LOW
BW	MEDIUM	HIGH	HIGH
SW	LOW	HIGH	HIGH

4.7   The duration of ballast water within an enclosed ballast tank will also be a factor in determining the number of surviving organisms. For example, even after 60 days, some organisms may remain in ballast water in a viable condition.

4.8   Because some aquatic organisms and pathogens that may exist in sediments carried by ships can survive to several months or longer, disposal of such sediment should be carefully managed and reported to port State Authorities.

4.9   In implementing ballast water and sediment discharge procedures, Port State Authorities should take account of all relevant factors.

5. IMPLEMENTATION

5.1   Member States applying ballast water and sediment discharge procedures should notify the Organization of specific requirements and provide to the Organization, for the information of other Member States and non-governmental organizations, copies of any regulations, standards or guidelines being applied.

5.2   Administrations and non-governmental shipping organizations should provide the widest possible distribution of information on ballast water and sediment discharge procedures being applied to shipping by port State Authorities. Failure to do so may lead to unnecessary delays for ships seeking entry to port States where ballast water and sediment discharge procedures are being applied.

5.3   In accordance with paragraph 5.2 above, ship operators and ships' crews should be familiar with the requirements of port State Authorities with respect to ballast water and sediment discharge procedures, including information that will be needed to obtain entry clearance. In this respect, masters should be made aware that penalties may be applied by port State Authorities for failure to comply with national requirements.

5.4   Member States and non-governmental organizations should provide to the Organization, for circulation, details of any research and development studies that they carry out with respect to the control of aquatic organisms and pathogens in ballast water and sediment found in ships.

5.5   Administrations are encouraged to report to the Organization incidences where compliance with ballast water and sediment discharge procedures required by port State Authorities has resulted in ship safety problems, unacceptably high costs, or delays to ships.

5.6   Member States should provide to the Organization details of annual compliance records for ballast water and sediment discharge procedures that they are applying. These records should report all incidences of non-compliance with regulations or guidelines and cite, by ship's name, official number and flag, all non-complying vessels.

5.7   Member States should notify the Organization of any local outbreaks or infectious diseases or water borne organisms that have been identified as a cause of concern to health and environmental authorities in other countries and for which ballast water or sediment discharges may be vectors of transmission. This information should be relayed by the Organization, without delay, to all Member States and non governmental organizations. Member States should ensure that problem species, endemic to their waters, are not being transferred from locally loaded ballast water. Masters of ships should be notified of the existence of problem species, including local outbreaks of phytoplankton blooms, and advised to exchange or treat their ballast water and sediment accordingly.

5.8   Member States should determine the environmental sensitivity of their waters to the extent deemed necessary. Ballast water and sediment discharge procedures should take into account the environmental sensitivity of these waters.

6. SHIP OPERATIONAL PROCEDURES

6.1   When loading ballast, every effort should be made to ensure that only clean ballast water is being taken on and that the uptake of sediment with the ballast water is minimized. Where practicable, ships should endeavour to avoid taking on ballast water in shallow water areas, or in the vicinity of dredging operations, to reduce the likelihood that the water will contain silt, which may harbour the cysts of unwanted aquatic organisms and pathogens, and to otherwise reduce the probability that unwanted aquatic organisms and pathogens are present in the water. Areas where there is a known outbreak or diseases communicable through ballast water, or in which phytoplankton blooms are occurring, should be avoided wherever practicable as a source or ballast.

6.2   When taking on ballast water, records of the dates, geographical locations, salinity and amount of ballast water taken on should be recorded in the ship's log-book. To enable monitoring by the Organization and port State Authorities, a report in the format shown in the appendix to these Guidelines should be completed by the ship's master and made available to the port State Authority. Procedures to be followed by the ship should be described in detail in the ship's operational manual. The sample used to determine the salinity of loaded ballast water should be obtained, wherever possible, from the ballast tanks themselves or from a supply piping tap. Surface sea water samples should not be taken as indicative of the water in the ballast tanks since sea water salinity may vary significantly with depth.

6.3   Subject to accessibility, all sources of sediment retention such as anchors, cables, chain lockers and suction wells should be cleaned routinely to reduce the possibility of spreading contamination.

7. STRATEGIES FOR PREVENTING THE INTRODUCTION OF UNWANTED AQUATIC ORGANISMS AND PATHOGENS FROM SHIP'S BALLAST WATER AND SEDIMENT DISCHARGES

7.1 General

7.1.1   In determining appropriate strategies for ballast water and sediment discharge procedures, the following criteria, *inter alia*, should be taken into account:

-    operational practicability;
-    effectiveness;
-    seafarer and ship safety;
-    environmental acceptability;
-    water and sediment control;
-    monitoring; and
-    cost effectiveness.

7.1.2 Approaches that may be effective in controlling the incidence and introduction of aquatic organisms and pathogens include:

-    the non-release or ballast water;

-    ballast water exchange and sediment removal at sea or in areas designated as acceptable for the purpose by the port State Authority;

-    ballast water management practices aimed at preventing or minimizing the uptake of contaminated water or sediment in ballasting and deballasting operations; and

-    discharge of ballast water into shore-based facilities for treatment or controlled disposal.

7.1.3 In considering which particular approach, or combination of approaches, to use, port State Authorities should have regard to the factors listed in paragraph 7.1.1.

7.2 Non-release of Ballast Water

The most effective means of preventing the introduction or unwanted aquatic organisms and pathogens from ships' ballast waters and sediments is to avoid, wherever possible, the discharge of ballast water.

7.3 Ballast Water Exchange and Sediment Removal

7.3.1 In the absence of more scientifically based means of control, exchange of ballast water in deep ocean areas or open seas currently offers a means of limiting the probability that fresh-water or coastal species will be transferred in ballast water. Responsibility for deciding on such action must rest with the master, taking into account prevailing safety, stability and structural factors and influences at the time.

7.3.2 Unlike coastal and estuarine waters that are rich in nutrients and life forms, deep ocean water or open seas contain few organisms. Those that do exist are unlikely to adapt readily to a new coastal or freshwater environment, hence the probability of transferring unwanted organisms through ballast water discharges can be greatly reduced by ocean or open sea ballast exchanges, preferably in water depths of 2,000 m or more. In those cases where ships do not encounter water depths of at least 2,000 m, exchange of ballast water should occur well clear of coastal and estuarine influences. There is evidence to suggest that, despite contact with water of high salinity, the cysts of some organisms can survive for protracted periods in the sediment within ballast tanks and elsewhere on a ship. Hence, where ballast water exchange is being used as a control measure, care should be taken to flush out ballast tanks, chain lockers and other locations where silt may accumulate, to dislodge and remove such accumulations, wherever practicable.

7.3.3 Care should also be taken when removing sediment deposits while a ship is in port or in coastal waters to ensure that the sediment is not disposed of directly into adjacent waters. Sediment should be removed to land-fill locations designated by the port State Authority or, alternatively, sterilized to kill all living organisms that it may contain prior to being discharged into local water bodies or otherwise disposed of.

7.3.4 Ships likely to be required to exchange ballast during a voyage should take into account the following requirements:

.1 stability to be maintained at all times to values not less than those recommended by the Organization (or required by the Administration);

.2 longitudinal stress values not to exceed those permitted by the ship's classification society with regard to prevailing sea conditions; and

.3 exchange of ballast in tanks or holds where significant structural loads may be generated by sloshing action in the partially filled tank or hold to be carried out in favourable sea and swell conditions such that the risk of structural damage is minimized.

7.3.5 Where the requirements of paragraph 7.3.4 cannot be met during an "at sea" exchange of ballast water, a "flow through" exchange of ballast water may be an acceptable alternative for those tanks. Procedures for exchange of this type should be approved by the Administration.

7.3.6 Where the requirements of paragraph 7.3.4 can be met during an "at sea" exchange of ballast water, before taking on exchange ballast water, tanks should be drained until pump suction is lost. This will minimize the likelihood of residual organism survival.

7.3.7 Where a port State Authority requires that an "at sea" exchange of ballast water be made, and, due to weather, sea conditions or operational impracticability, such action cannot be taken, the ship should report this fact to the port State Authority prior to entering its national waters, so that appropriate alternative action can be arranged.

7.3.8 Alternative action will also be necessary in those instances where ships may not leave a continental shelf during their voyage. Unless specific alternative instructions have been issued by a port State Authority applying ballast water and sediment controls, ships should report non-compliance prior to entering the port State's waters.

7.3.9 Port State Authorities applying ballast water exchange and sediment removal procedures may require ships to complete a ballast water control form or some other acceptable system of reporting. A model form for this purpose is in the appendix. Port State Authorities should arrange for such reporting forms to be distributed to ships, together with instructions for completion of the form and procedures for its return to the appropriate authorities.

7.3.10 In those cases where a ship arrives at a port without having carried out an "at sea" ballast water exchange, or has otherwise failed to carry out any alternative procedures acceptable to port State Authorities, the ship may be required to proceed to an approved location to carry out the necessary exchange, treat the ballast water *in situ*, seal the ballast tanks against discharge in the port State's waters, pump the ballast water to a shore reception facility, or prove, by laboratory analysis, that the ballast water is acceptable.

7.3.11 To facilitate administration of ballast water exchange and sediment removal procedures on board ships, a responsible officer familiar with those procedures should be appointed to maintain appropriate records and to ensure that all ballast water exchange and sediment removal procedures are followed and recorded. Written ballast water and sediment removal procedures should be included in the ship's operational manual.

7.3.12 Port State Authorities applying ballast water exchange and sediment discharge procedures may wish to monitor compliance with, and effectiveness of, their controls.

7.3.13 Effectiveness monitoring may also be undertaken by port State Authorities, by taking and analyzing ballast water and sediment samples from ships complying with prescribed exchange procedures, to test for the continued survival of unwanted aquatic organisms and pathogens.

7.3.14 Where ballast water or sediment sampling for compliance or effectiveness monitoring is being undertaken, port State Authorities should minimize delays to ships when taking such samples. Use of plankton nets, either by a vertical tow through ballasted deep tanks or cargo holds or by attachment to an open fire-main hydrant, suitably cross-connected to the ballast main, is one suggested means of ballast water sampling. Sediment samples may be taken from areas where sediment is most likely to accumulate, such as around outlet pipes, bulkhead and hold corners, etc., to the extent that these are accessible. Appropriate safety precautions must be employed wherever the taking of water or sediment samples requires tank entry.

7.3.15 Port State Authorities may also wish, subject to relevant safety considerations, to sample sediment in suction wells, chain lockers or other areas where sediment may accumulate.

7.3.16 In some cases, ships bound for ports which apply strategies for preventing the introduction of unwanted aquatic organisms and pathogens from ships' ballast water and sediments may avoid "at sea" exchange of ballast water, or other control procedures, by having their ballast water or harbour source samples analysed by a laboratory that is acceptable to the port State Authority. Where sampled and analysed ballast or harbour source water is found to be free from unwanted aquatic organisms or pathogens, an analyst's certificate, attesting to that fact, should be made available to port State Authorities. When analysis of ballast or harbour source water or sediment is being used as a control procedure, port State Authorities should provide Administrations with a target listing of unwanted aquatic organisms or pathogens.

7.3.17 Port State Authorities may sample or require samples to analyse ballast water and sediment before permitting a vessel to proceed to discharge its ballast water in environmentally sensitive locations. In the event that unwanted aquatic organisms or pathogens are found to be present in the samples, ships may be prohibited from discharging ballast or sediment, except to shore reception facilities or in designated marine areas.

7.4 Ballast Water Management Practices

7.4.1 Port State Authorities may allow the use of appropriate ballast water management practices, aimed at preventing or minimizing the uptake and discharge of contaminated water or sediment in ballasting and deballasting operations. Such practices may be used when adjudged as reducing the risks of introducing unwanted aquatic organisms and pathogens to a level acceptable to port State Authorities, who, for this purpose, may set conditions with which such practices need to comply.

7.4.2 Such conditions should include appropriate ballast water management plans, training of ships' officers and crew, and the nomination of key control personnel.

7.5 Shore Reception Facilities

7.5.1 Where adequate shore reception facilities exist, discharge of ship's ballast water in port into such facilities may provide an acceptable means of control. Port State Authorities utilizing this strategy should ensure that the discharged ballast water has been effectively treated before release. Any treatment used should itself be environmentally acceptable.

7.5.2 Reception facilities should be made available for the safe disposal of tank sediment when ships are undergoing repair or refit. Sediment, removed from ballast tanks and other areas of accumulation, should be disposed of in accordance with paragraph 7.3.3 above.

7.5.3 Member States should provide the Organization and ships with information on the locations, capacities and availability of, and any applicable fees relevant to, reception facilities being provided for the safe disposal of ballast water and removed sediment.

8. TRAINING, EDUCATION AND SHIPS' MANAGEMENT PLANS

8.1   Administrations and non-governmental shipping organizations should ensure that ships' crews are made aware of the ecological and health hazards posed by the indiscriminate loading and discharging of ballast water and of the need to maintain tanks and equipment, such as anchors, cables and hawse pipes, free from sediment.

8.2   Training curricula for ships' crews should include instruction on the application of ballast water and sediment discharge procedures, based upon the information contained in these Guidelines. Instruction should also be provided on the maintenance of log-book records, indicating the dates and times of ballast water loading, exchange or discharge, salinity and the geographical location where such operations are carried out.

8.3   Ships' crews should receive adequate instruction on the methods of ballast water and sediment discharge procedures being applied on their ship, including appropriate safety training in the relevant procedures.

8.4   Ballast water management plans should be incorporated into ships' operational manuals for the guidance of the ships' crews. Such plans should include, but not necessarily be limited to, information on the following:

- ballast water loading and discharging procedures and precautions;
- ballast water and sediment sampling and testing;
- controls applied by port State Authorities;
- reporting and information requirements;
- exchange and treatment options or requirements;
- crew safety guidelines;
- sediment disposal arrangements; and
- crew education and training.

8.5   Ships' operational manuals should include reference to these Guidelines and to the need to comply with any ballast water and sediment discharge procedures imposed by port State Authorities.

9. FUTURE CONSIDERATIONS

9.1   There is a clear need to research and develop revised and additional measures, particularly as new information on organisms and pathogens of concern becomes available. Areas for further research include, *inter alia*:

- treatment by chemicals and biocides;
- heat treatment;
- oxygen deprivation control;
- tank coatings;
- filters; and
- ultraviolet light disinfection.

It must be made clear, however, that there is a lack of research knowledge and practical experience on the cost, safety, effectiveness and environmental acceptability of these possible approaches. Any proposed chemical or biocidal treatments should be environmentally safe and in compliance with international conventions. Authorities carrying out or commissioning research studies into these or other relevant areas are encouraged to work co-operatively and provide information on the results to the Organization.

9.2   In the longer term and to the extent possible, changes in ship design may be warranted to prevent the introduction of unwanted aquatic organisms and pathogens from ships. For example, subdivision of tanks, piping arrangements and pumping procedures should be designed and constructed to minimize uptake and accumulation of sediment in ballast tanks.

9.3   Classification societies are urged to include provisions for ballast water and sediment discharge procedures in their rule requirements.

BALLAST WATER CONTROL REPORT FORM

(To be completed by ship's master prior to arrival and
provided to port State Authority upon request)

NAME OF SHIP: ....

PORT OF REGISTRY: ....

OFFICIAL NO. OR CALL SIGN: ....

OWNERS/OPERATORS: ....

AGENT: ....

IMO GUIDELINES CARRIED?   ☐ Yes   ☐ No

CONTROL ACTION TAKEN?- Non-release of ballast

        ☐ Ballast water exchange

        ☐ Ballast water management practices

        ☐ Use of shore reception facilities

        ☐ Other (specify) ....

        ....

        ☐ Nil

Information on Ballast Water Being Carried

Tank Location	Quantity (tons)	Geographic origin of carried ballast		Salinity of original sample (specific gravity)	Intended discharge port		If exchanged, where was ballast loaded?		Salinity of reballasted sample (specific gravity)	Controls used where ballast not exchanged
		Lat.	Long.		Place	Date	Lat.	Long.		
FOREPEAK										
AFT PEAK										
DOUBLE										
BOTTOM										
WING TANKS										
SIDE TANKS										
DEEP TANKS										
CARGO HOLDS										
OTHER (SPECIFY)										

MASTER'S NAME: ....

(PLEASE PRINT)

DATE: ....

MASTER'S SIGNATURE: ....

PORT LOCATION: ....